The Perfumier and the Stinkhorn

ALSO BY RICHARD MABEY

Weeds

Food for Free

The Unofficial Countryside

The Common Ground

The Flowering of Britain

Gilbert White

Home Country

Whistling in the Dark:
In Pursuit of the Nightingale

Flora Britannica

Selected Writing 1974–1999

Nature Cure

Fencing Paradise

Beechcombings

A Brush with Nature

THE PERFUMIER AND THE STINKHORN

RICHARD MABEY

P

PROFILE BOOKS

First published in Great Britain in 2011 by
PROFILE BOOKS LTD
3A Exmouth House
Pine Street
London EC1R 0JH
www.profilebooks.com

Shorter versions of these essays were originally broadcast by Richard Mabey on BBC Radio 3 in autumn 2009, in *The Essay* series, titled 'The Scientist and the Romantic'.

Illustrations by Michael Kirkwood

1 3 5 7 9 10 8 6 4 2

Printed and bound in Great Britain by
Clays, Bungay, Suffolk

A CIP catalogue record for this book is available from the British Library.

ISBN 978 1 84668 407 4
eISBN 978 1 84765 450 2

The paper this book is printed on is certified by the © 1996 Forest Stewardship Council A.C. (FSC). It is ancient-forest friendly. The printer holds FSC chain of custody SGS-COC-2061

Contents

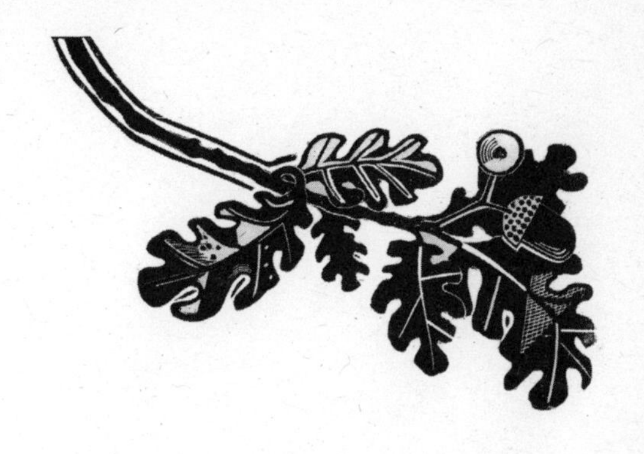

THE GREENHOUSE AND THE FIELD

1

The Greenhouse and the Field

WHEN I MOVED from the Chilterns to Norfolk in 2002, I took with me a longing to see a barn owl on my home patch again. They'd become scarce in Middle England, and I missed their pale vigils over the twilit fields. I'd been a year in my new patch before I heard about one. I was tipped off by our window cleaner, who glimpsed it most evenings when dog-walking. It proved to be a late riser, only materialising on the cusp of darkness. When I first saw it, the last light from the west was shining through its almost translucent wing tips. I became rhapsodic about the way it was 'shuttling night and day together', and scribbled notes about how it seemed to be 'winnowing the grass, threshing it for food'. I was burying the real bird – which would have rapidly starved if it had behaved like a threshing machine – under bushels of thoughtless visual metaphors. I could have done with some scientific ballast at that moment to ground my flights of fancy.

So when I'm occasionally called a 'Romantic naturalist' I wonder whether it's an accusation as much as a description: the meticulous observations of the natural scientist corrupted by my overheated imagination; objectivity compromised by my Romantic insistence on making feelings part of the equation.

Well, I suppose it depends what you mean by Romanticism. I rather incline towards Sam Coleridge and John Clare's view – that nature isn't a machine to be dispassionately dissected, but a community of which we, the observers, are inextricably part. And that our feelings about that community are a perfectly proper subject for reflection, because they shape our relationship with it – a more troubled relationship now than it ever was for the eighteenth-century Romantics.

In principle these ideals shouldn't conflict

with scientific rigour. Feelings can precede or follow the moment of exact observation without necessarily contaminating its truthfulness. But in practice marrying these two approaches is tricky work, and raises all kinds of puzzles about the terms of our experience of nature. Can you, for instance, closely observe a living organism without in some way taking it out of context, literally or perceptually? Can emotional engagement with nature amount to a kind of subtle take-over? Is it possible for us to sympathetically take another creature's sensory viewpoint without becoming anthropomorphic? Do the technological devices by which we enlarge our understanding of nature enhance or diminish our sense of kindredness with it? Running through these conundrums is the issue of the primacy of our senses, the only channels through which we can relate to the physical world. The natural scientist depends on them for information, but mistrusts their subjectivity and fallibility, and is chiefly interested in how they lead to *explanations* of nature. The Romantic revels in them for their own sake. They provide sensual experiences as well as sensory data, and are agencies we share with the rest of nature. Wolves and owls and bumblebees stare, sniff and listen too, and the Romantic wants to be part of that great global conversation.

In these six essays I want to reflect on my own rickety attempts to marry a Romantic view of the natural world with a mite of scientific precision. Each essay concentrates on a particular sense – sight, taste, smell, hearing, finding your place. In this essay in an oblique way, I'm thinking about touch, which is unique among the senses in being both passive and active, about feeling and manipulation.

I began to be fascinated by the natural world in the wasteland that lay at the back of our family house in the Chilterns. This quarter-mile square of unkempt grass and free-range trees had an exotic history, though I didn't know it at the time. It had been the grounds of a Georgian mansion owned by Graham Greene's uncle Charles, and, before the First World War, the young novelist-to-be was a frequent visitor. He had a secret eyrie on the

roof of the Hall, from which he'd gaze across the familial parkland and dream of being an explorer.

The estate was broken up in the 1920s, and the old park abandoned. By the time our neighbourhood gang occupied it in the 1950s it was a thrilling wilderness of feral trees, unplumbable wells and shoals of mysterious dells and mounds that were like the tumuli of a lost civilisation. We called it simply 'The Field', as if there wasn't another one worth bothering about, and used it as our local common. I saw my first barn owl there, hunting over what had once been the Hall's tennis courts. I learned how to make dens under the roots of fallen trees, how to walk painlessly over flints, and dreamed, just as Graham Greene had forty years before, of being an explorer. We had a Romantic life out there, playing at noble savages.

But at the edge of The Field I had another kind of den, where the business was more formal. My father's greenhouse backed onto the wasteland, and when I was about eleven he let me turn one end of it into a makeshift laboratory. As scientific establishments go, it must have been one of the strangest. I'd garnered all kinds of household chemicals from Mum's kitchen – washing soda, borax, vinegar, lime, Epsom salts – and put them in jam jars, carefully labelled with their scientific names. They sat in neat rows on the breeze blocks where Dad's flower-pots had been. In those days Boots sold chemicals and apparatus over the counter, even to kids, so my pocket money turned into phials of potassium permanganate and sulphur and a gleaming array of flasks, pipettes and funnels. But no dissection knives or collecting bottles. I was mad for chemistry then, but not biology, and it would never have occurred to me to take any living things out of The Field for investigation on the bench. They belonged out in that wild savanna beyond the fence.

When, much later, I began rationalising my memories of that time, I thought of the strand of barbed wire separating my lab from The Field as a symbolic boundary between science and romance, as decisive as the line between C. P. Snow's 'Two Cultures'. But now I realise it was nothing of the kind. I was up to the same business in both places. Out in The Field I may have been ravished by the wild luxuriance of it all, but I was also learning about nature by sheer physical experience – by touch. I began to appreciate the properties of different kinds of wood – which burned well, which were best for the roofs of dens, which bent into weapons. I succeeded in making a modest pat of butter by spinning a jar of cream on the wheel of an upturned bicycle. And I unwittingly did the first research for a book I was to write twenty years later, by learning to eat hawthorn leaves. They were weird, but they made a change from fire-blackened potatoes.

And just as I was a naturalist in the field, I was a Romantic in the lab. I was fond of the theoretical side of chemistry, and had begun to grasp the rudiments of the Periodic Table of elements (and memorise it: my dad would give me half a crown for a recital). But this wasn't the kind of thing I was up to in the lab. My end of the glasshouse was a magician's chamber, a theatre of metamorphosis, where I could wave a wand and revel in the sensuous transformations of matter. The orange crystals of potassium dichromate would produce lurid pigments when mixed with lead and zinc salts and various acids. Powdered sulphur turned into a gooey plastic when heated, and iron filings plus a dash of sulphuric acid released the uproarious rotten-eggs stench of hydrogen sulphide. As for the reactions of mercury, well, I shouldn't have been in possession of this toxic liquid metal at all. But Boots sold it, and I'd filched some from school. I was able to hide the heavy globules – as mobile as small ferrets – in my pocket. I even made some of my own apparatus, blowing small flasks and thistle funnels from glass tubing (Boots again) held over a Bunsen burner fuelled from the gas-socket in the kitchen.

My professionally made magic wands were more sophisticated, thanks to my dad, who was an insatiable collector of gadgets. They were often stuffed into his briefcase with bizarre bits of meat he picked up at Smithfield Market, and were a kind of technological offal themselves: eye-surgeons' scalpels, war-surplus dynamos, vast spools of copper wire … anything that would, as he always put it, 'come in useful one day'. It usually did. I once rigged up a bit of ramshackle apparatus for electrolysing water into its basic elements. I cut down an old glass aquarium to bowl size, and put in one electrode of copper, and another of zinc – the latter made from one of the flanges that held the greenhouse roof-glass in place. When they were covered with water, and wired up to my hand-cranked dynamo, they would fizz with fine streams of oxygen and hydrogen. I collected the gases in jars and burned them carefully together. They turned back into water – one of the reactions that underpins life itself.

For me the romance of these experiments wasn't just their spectacular revelation of the wonders of matter. *I* had made them happen. I'd discovered the ambivalent power of touch, the basis of all technology.

In my late teenage years the pull of The Field, and of my greenhouse laboratory began to wane. But when I went up to Oxford I opted to study biochemistry, in the hope that I might bring those two halves of my life together

– and promptly changed to philosophy when I saw the animal experiments I was expected to perform. Those were the years when the threat of nuclear war was at its height. Rachel Carson's prophetic book about pesticides, *Silent Spring*, had just been published, and the public image of science was at a low. And I'm sad to say that for a while I shared the widespread view of scientists as clones of Dr Frankenstein.

When I became a writer myself I felt that I wanted to explore our perceptions of nature as much as the natural world itself, and especially the sense of sympathy, and what it says about the experiences and needs common to all living things. But all too often it led to events like my Norfolk barn owl encounter, where nature's own agenda got buried under my own, with its pall of aesthetic and emotional metaphors.

I wish I'd become acquainted earlier with the nineteenth-century poet John Clare, who perfectly balanced an intense Romantic imagination with a sympathetic naturalist's eye. Clare is often thought to be an enemy of science, partly because he once called it

a 'dark system'. But, as he wrote himself, 'I puzzled over every thing in my hours of leisure that came my way. Mathematics Astronomy Botany and other things with a restless curiosity that was ever on the enquiry and never satisfied.' What he did object to was the oppression of living things in the name of science, whether by collecting and killing, or by obscuring them with arcane classification systems. Clare was always excited by finding new species of plant and insect, but insisted that he had 'no desire further to dry the plant or torture the butterfly by sticking it on a cork board with a pin'. Instead he'd prefer to watch the 'butterfly settle till he can come up with it to examine the powdered colour on its wings'.

Clare is credited with the first county record of sixty-five birds and more than forty plants purely from the compelling accuracy of his poetic descriptions. Part of this record is the extraordinary catalogue he made of local orchids. He disdained unpronounceable Linnaean names, which he believed took living things out of their own and our cultural context. His twelve orchid species are described by their vernacular names, and by their addresses. The butterfly orchid, for example, 'Grows in low part of Mr Clark's close at Royce wood end & … was very plentiful before Enclosure on a Spot called Parkers Moor near Peasfield-hedge and on Deadmoor near Sneef Green … but these places are now all under plough.'

A canny Linnaean might object that these homely descriptions are more human-centred than any Latinate name. But what Clare did was find a way of seeing and talking about nature in which humans and other organisms inhabit, equitably, the same spaces, and I'll be visiting him again.

THE LICHEN AND THE LENS

2

The Lichen and the Lens

WE TEND TO TAKE the astonishing power of our eyes for granted, until they go wrong, or their vision is, for some reason, mysteriously enlarged. This once happened for me ironically, plumb in the middle of an anxiety attack. I've been prone to these since I was a child, and the jolt they give to your normal attention can create a sense of unreality, of disjunction. You receive familiar images, but your concentration is somewhere else. It's often described as like having a pane of glass between you and the outside world. But it can be a lens, too. The wobble it gives to normal, unthinking perception can lead to a strange heightening of awareness. On this occasion

I was out walking, trying to concentrate on the landscape and not the weird feeling in my head, when I became convinced that I could make out the minute physical details of the world a quarter of a mile away – individual bricks, the ears of a man, discrete eddies in a plume of smoke, the wings of flying birds. It seemed, in that moment of hypersensitivity, to be some kind of supernatural gift; but of course I'd simply become aware of the sensory processing I did unconsciously every second of my life.

I'd acquired my first exterior lens much earlier. My parents had given me a pair of second-hand binoculars as a reward for passing the Common Entrance exam. For a schoolboy birder it was like getting your hands on a Philosopher's Stone, a magical instrument that could turn base spadgers into gold-finches. One afternoon I was walking a favourite beat in a valley close to my Chiltern home when I spotted what I thought was a robin, dashing flamboyantly up and down from a hedge. I focused my new equipment on it, and was transfixed to see that it had a swaggering chestnut-red tail and a jet-black face. I felt sure I knew what it was, and when I finally lost sight of it I raced home in a state of high excitement to look up 'redstart' in *British Birds in Colour* – the other half of my exam prize. There, in the thrush section, was John Gould's portrait of one in an alder tree, glowing the colour of red-hot iron, and my bird to the life. But the text puzzled me. It said the bird, a summer migrant, haunted ancient oak woodland. But there weren't any old oakwoods where I'd seen it. Had it been off course, or more excitingly, *on* course, bound for one of its upland strongholds? And using an ancestral flyway that took it from Africa through this narrow valley, *my* valley? If so, I'd found a point where our territorial paths crossed, and for a fleeting moment the world seemed, to my childish imagination, a more whole and comprehensible place.

When list-making palled later in my life, I began to relish the sheer surface beauty that binoculars could reveal – the blue sheen on swallows' backs, the lovely marbling of the gadwalls' plumage, the terrifying golden glare of a sparrowhawk's eye. But these revelations are always bought at a cost. The detail is heightened, but the context diminished. The bird framed in the circle of a lens is a creature partially abstracted from its habitat and its interaction with other birds. Not long ago, while I was in a boat on the Norfolk Broads, an osprey appeared and began flying over us along the dyke. It was stupendous against the vast sky. It dwarfed the marsh harriers that rose up to mob it and for a brief moment seemed to dim the sunlight. Its harassers strafed and retreated, then dashed recklessly in again. The osprey hovered, dipped, and finally went into a spectacular stoop into the water for a fish, followed by the harriers, which almost went in with it. So much was happening that I put down my binoculars and watched the whole dramatic scene unfold with my naked eyes.

At such moments technology can be a barrier and over-privilege its user, breaking any sense you have of being engaged with what you're watching. Of course, I'm being hopelessly inconsistent about this. Identifying that nomadic redstart and glimpsing what our encounter meant, couldn't have happened without my binoculars. The insights into

nature that technology can provide are almost always double edged.

In the late eighteenth century, before the Romantic Movement was really established, another optical gadget was fashionable. The Claude glass, named after the French painter Claude Lorrain, was a hand-held, slightly convex mirror, which provided on-the-spot, selective, ready-framed reflections of promising views. Significantly, the person using it usually stood with their *back* to the real landscape, shifting backwards and forwards, waving the mirror about until it contained a pleasing image. The business wasn't without its dangers. The poet Thomas Gray fell into a ditch whilst backing off to improve an image. And a much more tragic tumble became the focus for one of the defining moments of the Romantic era. Early in the nineteenth century, William Wordsworth climbed Helvellyn in the Lake District with the novelist Walter Scott and Humphry Davy, the brilliant chemist. He was leading them to the spot where the body of a young artist, Charles Gough, had been found after a calamitous fall a few years earlier – with a Claude glass about his person. It's not clear what these three prophets of the new age hoped to find on their pilgrimage. Perhaps the echo of a heroic death in the search for beauty, or the triumph of nature over art. But their account suggests that they felt the glass was implicated in some way, and were gripped by this symbol of the downfall of an over-regimented concept of 'natural' art.

The Claude glass was a product of the Picturesque movement, which taught that landscapes and natural scenes could be fully appreciated only when judged as if they were pictures – and that art could be refined only by more attentive study of landscape and nature. Painting and landscapes have been locked in this self-referential bind ever since, to the extent that pictures may now be our communal models for how scenery – and by implication nature – should look. The view from the top of the hill, the artfully framed scene in the Claude glass, the prospect that puts the viewer himself at the focal point, have come to dominate our visual perspective on the natural world. Even among painters who might loosely be described as Romantic, it is hard to find a view from the hedge-bottom or the *inside* of a wood.

But not in poetry. I mentioned in the first of these essays John Clare's highly visual take on the world, and his insistence that other organisms had perspectives too. He had a custom on his walks of what he called 'dropping down', which was both a way of peering closely at the earth, and the posture in which he scribbled his verse, on scraps of paper and old seed packets. In his poem 'To the Snipe' – addressed *to* the bird, out of sympathy and respect – he drops down comprehensively, and pictures the snipe's boggy habitat from *its* point of view, as a wilderness towering above the quagmire. The reeds are transmuted into giant trees: 'the clump of huge flag forest that thy haunts invest … suiting thy nature well'. Clare's ramping dialect catches the swampiness, the strangeness, the privacy of the bird's home: 'The little sinky foss / Streaking the moors whence spa-red water spews / From pudges fringed with moss.'

Clare's dropping down opens up a new perspective on the world. A swamp is usually seen as an undifferentiated morass of greens, which lies *under* our feet. Clare, imagining it through the snipe's senses, transforms it into a place of teeming ecological detail, spreading *over* us, and dwarfing us by its complexity.

This is what a microscope can do, dropping us down into small and insignificant worlds. I held off from using one till quite late in my life, fearing that it would reduce organisms I knew as living entities into inanimate fragments. But exactly the opposite happened. I acquired a stereoscopic microscope chiefly to look at lichens, which in their haunts on trees and rocks can look at times like no more than superficial ornaments. But magnified a hundred times they become labyrinths of complexity. And it isn't just pretty structural patterns you see but whole unexpected life processes. Diving down in three dimensions through the architecture of the plant it becomes clear that it's symbiotic, a partnership of *two* plants – a fungal shell impacted with the minute green cells of a food-producing algae. There are tiny insect eggs embedded in the fungus. And then, scurrying through the tangle of trunks and roots, is an insect itself, a grey, louse-like creature no bigger than the eye of a needle. It is *browsing* on the lichen. There are even microscopic toadstools growing on the fungal surfaces. There is an entire forest ecosystem in this one square centimetre. It is a fractal world, each magnified layer reflecting the structure and processes of the one before.

The American poet Gary Snyder has questioned the inevitability of looking at the natural world from a conventionally human-centred viewpoint. In his essay 'Unnatural Writing', he traces connections between our stories of the world and other creatures' 'narratives'. 'All our literatures are leavings', he writes, 'of the same order as the myths of wilderness people who leave behind only stories and few stone tools. Other orders of being have their own literature. Narrative in the deer world is a track of scents that is passed on from deer to deer, with an art of interpretation which is instinctive. A literature of bloodstains, a bit of piss, a whiff of oestrus, a hint of rut, a scrape on a sapling, and long gone.' Later in the essay he argues that conventional natural history and science writing are what he calls 'naively realistic', in that they unquestioningly view nature from the perspective of the front-mounted bifocal human eye.

We can't 'see' as a dragonfly does, of course – though with the help of computers we can reconstruct the images which its multiple-lensed eye presents to its brain. But we can subvert our orthodox view of nature, in which, either literally or culturally, we're invariably at the focal point.

Recently our view of one of the least known of the planet's habitats has been transformed by a simple change of perspective. The top storey of the rainforest used to be seen exactly in accordance with its conventional name – as a canopy. In a kind of mirror image of Clare's swamp, it was defined from our groundling's point of view as an undifferentiated parasol, a mere provider of shade for the important earthbound business below. The traditional way of exploring it had been

to hack the trees down – at which point, of course, its intricate ecosystem collapses.

But in the 1980s, a group of French biologists had the idea of approaching the canopy from above. They developed an inflatable raft in the shape of a starfish, which could be delicately lowered onto the canopy to provide a platform for the scientists. The vague overhead sunscreen was now reconfigured as highly detailed *ground*. And what was discovered there challenged many of our other assumptions about the nature of life on earth. It looks, for example, as if more than half of all earth's plant and animal species live in the canopy; that most of the plants don't need fertile soil, but live on rainwater and sunlight; and that partnership, or symbiosis, is the norm.

The most extraordinary – and most powerfully Romantic – image of our lifetimes is the unforgettable portrait of our planet from space. The earth beneath our feet, cast as our property, taken for granted, riven into myriad disconnected systems, was suddenly glimpsed as something beyond us – a single place, fertile, vulnerable, terribly alone. The American essayist Lewis Thomas wrote: 'Viewed from the distance of the moon, the astonishing thing about the Earth, catching the breath, is that it is alive. The photographs show the dry, pounded surface of the moon in the foreground, dead as old bone. Aloft, floating free beneath the moist, gleaming membrane of the bright blue sky, is the rising earth, the only exuberant thing in this part of the cosmos.' A framed vision, provided by technology, that transformed our whole cultural frame of reference.

THE CRAB APPLE AND THE GRAFTING KNIFE

3

The Crab Apple and the Grafting Knife

TASTE IS NOT QUITE LIKE other sense experiences. It's complex, synergistic, a bringing together of the sensations of a whole range of organs. It might more helpfully be called 'the experience in the mouth'. The lips and tongue (equipped with the same kind of nerve-endings as the genitals) register 'mouth-touch' – surface texture, elasticity, bite. More specialised receptors in the tongue (we call them 'tastebuds', with the pleasant suggestion that they can be 'opened' by the right titillation) pick up the elemental qualities of sweetness, sourness and saltiness. But what we commonly call taste is really flavour, which is synonymous with scent, and that is something for the next essay.

Scents – unless they're powerful, or disgusting – don't catch your imagination when you're young. They're not experienced separately from the overwhelmingly seductive business of eating. I was chewing nature down long before I thought of sniffing it, and those childhood experiments with hawthorn leaves and raw chestnuts and impossible-to-digest blades of grass left a kind of aftertaste that was rekindled a decade later.

In my late teens I'd begun spending weekends and holidays on the north Norfolk coast with a group of friends. We bedded down in a converted lifeboat moored in Blakeney harbour, and spent our days playing at being bohemians, birdwatching, sailing and running wild on the vast, mutable saltmarshes. It was all very new, and I was fascinated by the practical view the locals had of a landscape that seemed so transcendental to me. They were still part-time hunter-gatherers. They went out cockling and winkling, diverted the local bus when big mushroom flushes appeared on the grazing marshes, and most strangely, picked wild vegetables from the creeks and sea-walls: sea-spinach, fennel and especially samphire, whose bright green, succulent shoots – they seemed half-seaweed, half-thornless cactus – grew as densely as grass on the intertidal mudflats. Samphire is often called poor man's asparagus, but I found eating it more like gulping down a sea-breeze, full of iron and ozone and intimations of thirstiness.

Eating wild food seemed a wonderfully direct way to get close to the natural world – a way, so to speak, of incorporating it. It also had the kind of higgledy-piggledy mix of science and culture that I'd been in thrall to since my first teenage dreams of writing. There was the prospect of Romanticism in the kitchen and of scholarliness in the hunt, and it wasn't long before I had the idea for my first real book, *Food for Free*. I'd explore the tradition of foraging in Britain, and see how it might be practically revived.

The research was bliss. I trawled through old books, and not just herbals and ancient cookery manuals. Wildings poked through everywhere – in fiction, poetry, scientific papers – and I hunted them like a literary scavenger. I raked through John Evelyn's waspish vegetarian tract *Acetaria. A Discourse on Salletts* (1699), through learned dissertations on the stomach contents of mummified Neolithic corpses, through Eric Linklater's classic novel *Poet's Pub*, with its improbable recipe for a roasted crab-apple and beer cup, called Lamb's Wool.

I found the origins of the fungus foray among the outings of the Woolhope Field Club in Victorian Herefordshire. Their *Proceedings* for the autumn of 1869 describe a foraging expedition by thirty-five of their members (nine of them vicars). They ranged around the local woods and fields by carriage, stopping off at likely hunting grounds, measuring fairy rings and gathering a hoard of edible mushrooms: milk-caps, ceps, chanterelles, wood hedgehogs, parasols. The day ended in the Green Dragon in Hereford, with the exhibits strewn out on the pub tables and a late lunch of the day's best trophies.

During World War II, foraging became both patriotic and necessary. Vicomte de

Mauduit's splendidly titled broadside *They Can't Ration These* (1940) was joined by the Ministry of Food's own pamphlet, *Hedgerow Harvest* which moved the Home Front out into the wild, with recipes for the obligatory Vitamin C-rich rose-hip syrup and sloe-and-marrow jam: 'If possible crack some of the stones and add to the preserve before boiling to give a nutty flavour.' What a Romantic exhortation for a country with its back to the wall!

But it was Dorothy Hartley who was my hero, and chief inspiration. Working as a pioneering female journalist, she had toured the countryside by bike, sleeping in ditches and collecting traditional recipes from remote farms. Her quirky masterpiece, *Food in England*, was published in 1954, and is full of historical and sensual gems: stories of the Kentish hop-pickers' way of cooking hop trimmings; an ethereal blackberry junket made simply by leaving the strained juice in a warm room; a glimpse of chanterelle mushrooms, 'sometimes clustered so close that they look like a torn golden shawl down among the dead leaves'.

What these accounts and enthusings said to me, in the idealistic mood of the 1960s, was that foraging could put you back in touch with the basic roots of food, with a world of lost scents and flavours. The hunt sharpened your senses, your whole awareness of how landscape and season and vegetation were interconnected. In my field research I resolved to follow the example of the doughty Ms Hartley, and reasoned that anything that hadn't been specifically identified as toxic (I clung to the Ministry of Agriculture's official handbook on British poisonous plants like a life-raft) was worth a go. Working on a presumption of innocence until taste proved otherwise, I experimented with everything from acorns and dock leaves to the bulbous galls on a thistle – which, before I bit into their disgustingly acrid flesh, reminded me of small kohl-rabi. That was a bad experience, but there were revelations. The new-potato savour of young burdock stalks; the bloomed drupes and bursting juiciness of dewberries – miniature grapes on cocktail sticks; the delectable marshmallowness of giant puffball, which always seemed inextricable from discovering the great white mounds, like soft standing stones or vegetable squids, at the corner of a pasture.

I was living part-time on the Norfolk coast by now, and trying to fulfil my dream of being a literary hunter-gatherer. For neatly contained spells I would flit from the Chilterns to the coast, and spend my days prowling the marshes like a scholar gipsy, binoculars over one shoulder and a bunch of sea-kale over the other. Back in my cottage I'd lay a formal place for one, clean tablecloth included, and eat the weeds of the day, washed down by a bottle from the Wine Society. Going feral was clearly not what I was after. I was clinging, for better or worse, to that hybrid course I'd mapped out when I was twelve, hunched in my makeshift lab with my bottles of kitchen chemicals, and gazing out at the wilderness just beyond the fence.

But it worked. I began to learn about plants. I started to understand how they worked, where they grew, what conditions they needed, how weather influenced their fruiting. The heady thrill of finding a crop, of experiencing bizarre new tastes, whetted new appetites and new sensitivities. 'Search *inside* a hazel bush for nuts', I wrote in my notebook, 'and scan them with the sun behind you, so that you can glimpse the nuts against the light.' And the taste of what I found seemed enhanced by this intimacy. Years later I read *Morel Tales* by the American sociologist Gary Alan Fines, about the 'culture' of wild mushroom hunting in the United States. His interviewees talked about the quality of 'gatheredness' that makes wild foods taste different from shop-bought ones, and about the ecstasy of discovery: 'Suddenly it is there in the shadows. A single, exquisite morel … stands by itself, boldly etched against the edge of the orchard. Awestruck at first, I am afraid to remove it. Perhaps it is the last morel in the world.' Henry Thoreau said much the same a century and a half before: 'The bitter-sweet of a white-oak acorn which you nibble in a bleak November walk over the tawny earth is more to me than a slice of imported pine-apple.'

*

A decade or so after I'd published *Food for Free* I met a well-known portrait photographer at a party. 'Ah, you're the man who eats weeds,' he pronounced, eyeing me as if I were an interestingly gnarled variety of turnip. 'What an interestingly *earthy* face.' This seemed to me to take the idea that you are what you eat a tad too far, and heightened the doubts I was beginning to have about the foraging revival. It seemed to be becoming too commercial, too ecologically careless, and to be increasingly driven by macho fantasies about 'surviving in the wild'. That was never what I'd been interested in, or thought possible or desirable in Britain.

These days I'm more of a wayside nibbler than a full-blooded forager. I grab single wild blackcurrants, overhanging the water as we nudge our boat through the Norfolk Broads. I like the astringent, aniseed taste of a few sweet cicely seeds, picked as an aperitif on a walk before supper. One late summer I found a bough from a roadside damson bush that had been flailed off while it was still in fruit, and experienced the improbable taste of a handful of sun-dried English prunes. The

1930s fruit gourmet Edward Bunyan used a phrase which perfectly catches the delights of this kind of casual garnering. Reflecting on evening meanders through his gooseberry patch he talked of the pleasures of 'ambulant consumption ... The freedom of the bush should be given to all visitors.'

The freedom of the bush: what a liberty! Not something to be indulged without a sense of responsibility. A couple of years ago in the lane behind our house in Norfolk, I came across what looked like fruit road-kill. Windthrown cherry-plums, yellow and scarlet, were scattered all over the road and being inexorably squashed by the traffic. They're rare fruiters here, and almost without thinking, I crouched down on the tarmac amidst the puddles and trash, and began stuffing them into my pockets. And in one of those insights that gathering your own food ought to bring you, it occurred to me that this is how many of the planet's citizens, of all species, find their daily rations. Working the margins. Making do.

My own most memorable times of ambulant consumption have been to do with hunting wilding apples, happenstance fruits that have sprung from thrown-away cores and bird droppings. Their randomness seems to catch all that is best about foraging: serendipity; a sharpness of taste, and of moment; a sense of possibility. I can still recall the trees along a Chiltern green lane that was my apple Elysian Way: a tree sniffed out from fifty yards away, lemon yellow fruit, scent of quince, too hard and acid to eat raw but sensational roasted; another with the bitter-sweet effervescence of sherbet; a third with long pear-shaped fruits and a warm smoky flavour behind the sharpness, as if they had already been cooked. They seemed like time capsules, echoes not just of lost orchards themselves, but of the *hortus conclusus* as a symbol of the proper relationship between humans and nature.

Thoreau loved wildings, and in an essay entitled *Wild Apples* he contrives a Romantic fantasy of varieties defined and named not by their botanical qualities, but by where they grow, the moments when they're picked, the mood of the picker: 'the Truant's Apple (Malus cessatoris), which no boy will ever go by without knocking off some, however late it may be; the Saunterer's Apple – you must lose yourself before you can find the way to that; the Beauty-of-the-air (Malus decus-aeris) ... the Railroad Apple, which perhaps came from a core thrown out of the cars; the apple whose fruit we tasted in our youth; our Particular Apple, not to be found in any catalogue, Malus pedestrium-solatium; also the apple where hangs the forgotten scythe.'

Thoreau's genius here is to see in our common experience of the powerful combination of place, time and taste an echo of the biological diversity of the world's apples. And this diversity is itself is a remarkable conjunction of wild nature and science. All the 20,000 varieties of apple that have been developed over the past four millennia spring from a single wild species, *Malus pumila*, a tree with very variable fruits which grows in the forests of central Asia. This aboriginal apple, in all its naturally occurring variety, was carried west by human migrants and foraging animals, each of them continually selecting the biggest and sweetest fruits. Then, about 4000 years ago in Babylon, grafting was discovered, perhaps from the chance observation

of the way chafing branches can fuse. From that moment it became possible to perpetuate the fugitive and unpredictable forms of *M. pumila* by implanting their branches on other apple trees. The technology of the grafting knife had ensured the survival of the myriad forms of the 'Beauty-of-the-air'. And the lingering of the offspring of these varieties by the railroad and the barn means the Romantic forager's territory becomes a kind of genetic reserve.

THE PERFUMIER AND THE STINKHORN

4

The Perfumier and the Stinkhorn

SMELL IS A VERY PRIMITIVE SENSE, but we only come to appreciate its subtleties in maturity. When I was a boy, despite my pretensions at being a young naturalist, I needed scents to be outrageous to be memorable. The one that broke through most gloriously was the stinkhorn fungus, its reek of rotting animal unmistakable at twenty yards.

But smells, unlike sights, are hard to describe. They inhabit an evocative, ephemeral space in our imaginations that is difficult to put into words. They can only be described by comparison with other smells. To be fixed in our imaginations they need to be attached to other memories – of place, moment, feeling – and that needs the experience of age.

As Marcel Proust, poet laureate of scent and memory, famously wrote: fragrance and flavour 'bear unflinchingly, in the tiny and almost impalpable drop of their essence, the vast structure of recollection'.

I can remember the last throes of the great drought of 1976, and how after three months in the sun we were all as hard-baked as the riverbeds. There was no sap left in anything. The air was scentless – except for that flintiness you sense close to a hot brick wall. Then, as abruptly as it had begun, the drought dissolved. September brought torrential rain. It made the pavements steam. It flashed off the parched earth and rushed straight back to the rivers. But some soaked into the ground, enough to start the cycles again. It was an ancient recipe: 'Dried earth. Just add water'. The next morning when I opened the window the air was full of an extraordinary fragrance that seemed to spark and capture every good memory I had. Seawater drying on skin. The chanterelles my friend Richard used to send me from Scotland in July, packed in moss. My dad painting the garden fence with creosote in autumn. A whiff of burned rubber on a

summer street. All these deeply personal and emotional recollections from the simple exhalation that follows a shower of rain on dry earth.

But it's a real enough aroma. It's called 'petrichor', the essence of stone, and it is all the things that you imagine it to be. From fallen flower petals, flakes of oak-moss, pollens and resins and desiccated mushrooms, a huge ensemble of perfumed essences is washed into the ground and absorbed by porous stones and clay. When warm rain falls again, they're released back into the air to rekindle our memories of their ingredients.

The puzzle is why we are all still so *good* at scents, despite their having little relevance to our survival, and why they are so linked with emotion. These days I find that I partly navigate my way through the year by scents, and that they unlock memories that I sometimes didn't know I had, my own 'vast structure of recollection'.

The first fragrant wild flower of the year is spurge laurel, with tiny green flowers in late January that smell, as many early spring blooms do, of snow and honey, a lure for early insects that still hints of the torpor of the hibernating months. But my real trigger is moschatel, in March. The first time I sniffed its diminutive five-faced flowers I was astonished to smell my first girlfriend. As a naïve sixteen-year-old I never knew what to make of this earthy musk and almond aroma on such an unearthly creature, and assumed it was some expensive perfume. Now, probably only slightly wiser, moschatel still transports me to those breathless clinches with Liz, and leaves me wondering about the strange evolutionary pathways that link the scent of a spring flower with a pubescent girl. Bluebells are even more sultry. In small numbers, I love the spicy, liliaceous overtone in their scent. But en masse, I find them almost narcotic, a heavy haze that catches the throat – or maybe, if I'm honest, makes me feel too nostalgically smothered by memories of a childhood spent in the woods. Then there is the scent of burnet roses, like warm cream, irrevocably linked for me with the limestone hills of the Burren, and my dear late friend Tony Evans, taking his exquisite photograph of their petals floating in a rock-pool.

Summer's smells are less emotionally intense, the perfumes of pure pleasure. The scent of crushed wormwood, or any member of the artemisia family, takes me in a flash to the north-Norfolk saltmarshes, where the bracing tang of sea-wormwood blends with the wind and the calls of curlews, and where I first had a house of my own. Meadowsweet is the signature scent of my new home in south Norfolk, a ubiquitous, astringent fragrance that rises up wherever you walk in a damp place, whether it's a fen or roadside ditch. It's two scents, really: the fishy, sexy, may-like sweetness of the flowers, and the almost medicinal cucumber-and-carbolic of the leaves. Maybe that's why one of its vernacular names is 'Courtship and Matrimony'.

And then there's gorse, or furze, or fuzz, which is in bloom somewhere every month of the year. 'When gorse is in blossom kissing's in season', the old saying goes. Its coconut and peach perfume has been a companion throughout my life, on heaths in the Chilterns and East Anglia, as a background to my first thrilling nightingale songs, as a florid savour in experimental wild salads. But I've had another kind of reaction to gorse, one that has nothing to do with memories. It's a June afternoon, and I'm lying on my back on a patch of heath. The gorse and broom flower-sprays are hanging like garlands against the sky. The smell is tropical – vanilla and melon are there along with coconut. Out of the blue, I'm hit by an extra burst of scent. It seems to fill not just my nose, but my eyes and cheeks. I wonder for a moment if a breeze has got up, but it's a day of dead calm. A few minutes later it happens again, and I recall noticing these rhythmic gusts with other kinds of flower – lilacs, lime blossom, viburnums. I wonder if it's a kind of olfactory illusion, a momentary dulling of attention. Something in me, not the plant. Violets do this because of a chemical called ionone, which briefly anaesthetises our scent buds. The flower itself continues to smell but we lose the ability to register it. Then a couple of minutes later, the nose recovers and the scent reappears. Or is it that some plants budget their precious scent molecules, emitting concentrated puffs as comeons to insects? Then I have a more outlandish thought. Does the gorse smell *me*, and know

there is a living thing near it? Is it directing its fragrant come-ons *my* way?

This was an outrageously egocentric notion, but not out of the question. Natural smells are not just random chemical emissions. They're part of a complex messaging system between plant and plant, animal and plant. Rats emit an airborne chemical signal – a pheromone – when they're afraid, which turns on a natural analgesic in other rats in the vicinity, to prepare them for pain. When oak leaves are seriously munched by insects they emit another pheromone, which promotes the production of extra tannin in neighbouring trees, and makes their leaves more bitter to marauders. Mopane trees in Africa, a favourite food of elephants, do the same, and send out warning messages to other trees when they're being browsed. The elephants are wise to this trick. They eat only a few leaves from each tree, and move upwind to new trees. 'We can't hear the trees calling to each other,' wrote Colin Tudge, 'but the air is abuzz with their conversations none the less, conducted in vaporous chemistry.'

The reason we know so much about scents

that we ourselves can't smell is thanks to a piece of technology invented by James Lovelock, proponent of the Gaia hypothesis. This suggests that communication between all things on earth is so interconnected that the planet itself is tantamount to a single organism. In the 1960s Lovelock developed an instrument called the electron capture detector, which is able to detect minute traces of chemical, especially in the air. He has one in his home, where he practices science as – in his words – 'a cottage industry'. I visited him there once, and he warned me that a puff of deodorant at the other end of the house could ruin the readings.

It's been this instrument which has revealed that fruit flies will respond to as little as one hundred-millionth of a gram of pheromone produced by Cassia plants, and that lima beans affected by spider mite give off a volatile chemical in minute concentrations that attracts *another* species of predatory mite that feeds on the original mites. It has helped to untangle the extraordinary lifecycle of the large blue butterfly, whose larvae feed on wild thyme and produce honey from their abdomens. They also generate a pheromone that mimics the scent of ant grubs. The adult ants are attracted to this scent, gather up the butterfly larvae and take them back to their nests, looking after them as if they were their own offspring. All the while the larvae are singing to the ants, echoing the rhythmic noises of their grubs. Electron capture detection is also helping that gravely threatened creature, the bee. Honey-bees are able to read and interpret the chemical cues diffused into the atmosphere over a range of forty square kilometres, and convey the information back to other bees from their colony. But we now know that residues in the exhausts of cars using lead-free petrol react with the odour molecules from flowers, making them indecipherable to bees. This may be one of the causes of the now widespread problem of sudden hive collapse.

Smell isn't the oldest sense. The earliest cells must have first acquired an ability to orientate themselves in space and respond to warmth. But the identification of food and the necessity of interacting with other organisms entailed the development of this chemical messaging system, and we've inherited it. Long before we began to register scents consciously our behaviour was being guided by them. They helped with finding a mate and bonding with children and tribe, with locating food and avoiding danger, with interpreting the weather and the comings and goings of other creatures. The smell receptors were the foundations of the limbic system, a primitive centre concerned with basic emotions and the recording of sensation, and it was round this that the apparatus of memory began to evolve. Our brains are outgrowths of our noses. No wonder that smells remain the great carriers and triggers of potent memories. Smell and memory are both processed in the same ancient areas of our brains.

In Patrick Susskind's black and nose-gripping fantasy *Perfume*, the hero is a man born with no personal odour whatsoever. But he has an exquisite sense of smell, which leads him to become a supreme concoctor of perfumes. His lack of any personal scent disturbs other people, so he devises fragrances to make him socially acceptable – an 'odour of inconspicuousness', a hint of new-born baby, and eventually a scent so irresistible that it attracts a mob, who eat him alive in an ecstatic frenzy.

We're still part of the planet's buzzing chemical conversation. Although sight is now overwhelmingly our most important sense, we have five million scent receptors in our noses and three thousand genes encoded for smells, as against just three for perception of colour. And though it matters not a jot to our survival, we can still sniff the difference, with our eyes shut, between gin and whisky, roses and lilac, curry and stew, bluebells and balsam. The great American biologist and essayist Lewis Thomas had a vision of the entire planet self-regulated by its smells: 'In this immense organism, chemical signs might serve the function of global hormones, keeping balance and symmetry in the operation of various interrelated working parts, informing tissues in the vegetation of the Alps about the state of eels in the Sargasso Sea, by long, interminable relays of interconnected messages between all kind of other creatures.'

And we're informed too, still kept in touch with our love affairs and heydays by the scents carried on the wind.

THE NIGHTINGALE AND THE SONOGRAM

5

The Nightingale and the Sonogram

THE SONGS OF THE NATURAL WORLD have a special appeal to the Romantic, because they seem to imply that other creatures can be creative artists, too. Birdsong has long been thought an exemplar of 'natural' – maybe even heavenly – music. In an early fifteenth-century verse, 'The Flower and the Leaf', the nightingale is described as 'the rural poet of the melody'. In 1737, the popular natural history painter Eleazar Albin described song birds as 'sweet choristers of the woods' designed specifically by God to delight Mankind – though he then rather spoils his blandishments by giving instructions on how to catch and cage them. Eighty years on, John Bingley went one stage further and drew up a league table, a musical Top Ten, of our native songsters. He awarded marks out of twenty for five qualities: 'mellowness of tone', 'sprightliness', 'plaintiveness', 'compass' and 'execution'. The nightingale won hands down in almost every category, though the blackbird, mysteriously, won only four points for mellowness and nothing at all for plaintiveness – showing just how subjective and changeable aesthetic taste can be.

But for the Romantic with more than a smidgeon of scientific self-respect, this is where caution, or at least some real puzzles, begin to appear. What does it mean to say that one bird is a better singer than another? And by which species' standards – bird or human? What has aesthetics to do anyway with an activity that biologists tell us is nothing more than a means of attracting a mate and declaring territorial rights? And why should we, a species only anciently related to the birds, find their songs so pleasing, and sometimes as piercingly evocative as smells? I always feel *included* in a songbird's audience, from an obstinate hunch that there is something in common between what the bird is trying to convey and what I am feeling, some ancient resonance between two cortexes.

The similarities between birdsong and music are so obvious that it's not unreasonable to assume some deep link. It may be far-fetched to think that it's genetic – though birds and ourselves spring from common ancestors. But an ancient cultural mimicry, or inspiration, has to be on the cards. Birds sing at the same critical moments as early humans performed their own rituals: in the spring, at night and at times of the changing of weather.

Some of links with human music have been quite eerie. Mozart had a pet starling, which famously learned a theme from his G Major Piano Concerto – except that he changed the G natural to a G sharp, catapulting the whole phrase into the twentieth century. Large numbers of species have been heard transposing their songs into different keys, that is, keeping the intervals between the notes the same but pitching the entire phrase higher or lower, in strict accord with the Western tonic sol-fa.

But it's the nightingale which has always

been regarded as the greatest virtuoso. Its musicality and apparent passion has made it the most versified bird in Western culture. The raw ingredients of its song might not impress you, but neither would the individual notes in a Beethoven symphony. The nightingale's phrases are very diverse, and some birds have many hundreds in their repertoires, but they are always unmistakably those of a nightingale. The full song may use phrases mimicked from thrushes or robins, but it will never sound like the song of any other bird: creativity in the bird world has its limits.

It often begins with a flourish of piping notes, sometimes in triplets or short arpeggios, then a single note repeated as a rich liquid bubbling The most extreme version of this last phrase is so distinctive as to be the bird's signature sound. It's a sequence of long, sighing, indrawn whistles, mounting in crescendo on a single note until they fall, in sheer bathos, to a dull rattle. Yet the variations on these phrases and the way they are put together are almost limitless.

For me, the quality in the nightingale's song that is absent from other birds' is *oratory*.

Its crescendos and silences, redolent with anticipation, make it hover on the edge of opera, or recitative. The most affecting nightingale performance I've heard was in Suffolk, in mid-May. The setting was narcotic – a full moon, mounds of cow parsley glowing like suspended balls of mist, a dark wood arching like a lustrous whaleback across the whole span of the southern horizon. And a nightingale started up just a few feet away from me. It was a shaman, experienced, rhetorical, insistent, and I sank into its charms, a willing initiate. A shooting star arced over the bush in which it was singing. As I edged closer, I became aware that my peripheral vision was closing down, and that I had no sense of where I was in space. And then, for just a few seconds, the bird was in my head, and it was me that was singing.

Where does this compelling, mystifying intensity come from? The song enchants most people who are prepared to listen, but makes us ask questions too, about what the bird is doing, how it is making choices about what to sing, whether it has some instinctive sense analogous to our own feelings for musicality.

Over the past 2000 years poets, scientists and musicians have tried to puzzle out what the nightingale is saying. For medieval troubadours, nightingales were fellow singer-songwriters, and often rather better at it than they were. In the wonderfully funny thirteenth-century poem *The Owl and the Nightingale*, the bird's sprightly song stands for the emerging ideas of free-will and individualism, against the ascetic philosophy of traditional Christianity, represented by the owl.

For much of the sixteenth and seventeenth centuries nightingale verse was maudlin, interpreting the sobbing notes in the song as symbolic of failed or hopeless love. (Ironically, biologists have half-confirmed this recently, discovering that the most vehement late-season singers are males who have failed to find a mate.) Coleridge was the first of the Romantics to blow away these melancholy associations, insisting that the song was joyful, a hymn to life, and, correctly, that it was a love-chant by the male bird. Poetic nightingales before this had invariably been female. Even the fastidious John Clare mistook the sex of the singing bird in his great poem *The Nightingale's Nest*. But he accomplished something no poet or scientist had attempted before: a note-by-note transcription of the sounds of a nightingale singing in his orchard. Eschewing any musical or poetic references he simply attempts to catch the sound through a lexicon of some twenty-odd sonic phrases: wevy wit, pelew pelew, jug jug jug. Of course, this being Clare, it turns into a kind of concrete poem. But it's also as scientifically objective as a modern sonogram, which registers notes and noise graphically.

In the late nineteenth century, George Meredith wrote a profoundly Gaian poem called *Night of Frost in May* in which he describes the nightingale as 'the lyre of earth ... it holds me linked; /Across the years to dead-ebb shores / I stand on, my blood-thrill restores.' In the twentieth century, the idea of the nightingale as a concert performer returned. In 1924 the cellist Beatrice Harrison performed an historic duet with a bird in her Surrey garden, which was broadcast live on the BBC's Home Service to more than one million listeners. In the 1930s a pair nested, unusually, in Derbyshire, and a local bus company laid on a 'Nightingale Special', ferrying passengers to the wood at sixpence a head. 'Short tough colliers,' one commentator wrote, 'crowded along the lane where the nightingale sang … and listened as earnestly and critically as they did when the parish choir performed *The Messiah*.'

But this perception of musicality is necessarily filtered through our own limited hearing apparatus and cultural assumptions. What might a nightingale's song sound like to another nightingale? Birds' sense of hearing is much more acute and discriminating than ours, and their songs compress sound information at pitches and speeds which are beyond the scope of human ears. My friend the composer David Hindley has discovered the true extent of the difference by slowing down and making meticulous computer-assisted analyses of birdsong. It has been a revelation. What sound to us like single notes resolve, at quarter speed, into complex chords. The sonograms are dense and massy. Birds' voiceboxes are capable of producing at least four notes simultaneously, including maybe a bass drone and some high percussive

clicks. A nightingale song, slowed down, is barely recognisable. It becomes an unsettling and unfamiliar recital of modernist effects – slides, whoops, rattles. Hindley likens the sound to early Webern.

Why should this be so? Wouldn't high scores in Bingley's categories of 'compass' or 'execution' be sufficient to demonstrate the male bird's genetic fitness and ability to hold territory? Perhaps female birds are impressed by complexity. Or maybe the singer himself is. The distinguished ornithologist Charles Hartshorne argues that the singing male himself is the first (and sometimes only) audience for his performance: 'be it noted,' he writes, 'that maintaining a territory – and still more attracting a mate – is only intermittently a desperate, all-or-none affair like keeping possession of a meat-bone … There is seldom an extreme emergency … It is not to be imagined that a bird engaged in territorial singing for hours is exclusively reacting to the possibility of successful invasion. There is ample room, and some probable need, for its activity to be sustained by the interest of the activity itself, both as a muscular exercise and as the production of a pattern of sound of which the bird itself is aware.'

Richard Dawkins agrees, and makes an intriguing additional argument in his book *Unweaving the Rainbow.* Taking his cue from the phrase in Keats' 'Ode' – 'and a drowsy numbness pains / My sense as though of hemlock I had drunk' – he suggests that the idea of a nightingale song working like a drug isn't entirely far-fetched. Some ornithologists think of the song as conveying precise information about the male birds' breeding condition and the like. 'But', argues Dawkins, 'another way to look at it has always seemed to me more vivid. The song is not informing the female but *manipulating* her. It is not so much changing what the female knows as directly changing the physiological state of her brain. It is acting like a drug.' And, to complete the circuit, it isn't that surprising that it also manipulated the mind of John Keats, and by extension, any human listener. We share the same fundamental vertebrate nervous systems as birds.

I like this argument, but find it a little narrow, and propose a compromise explanation,

based on the available evidence. Birdsong may not be a language, but it is expressive – expression*ist*, if you like. It conveys a bird's emotional tone, be it proprietorial, angry, sexy, contented, sociable, exuberant – states of mind we intuitively understand.

Even the commonest birds work this spell. Much more than the nightingale, or the increasingly scarce swallow, it's the blackbird, I suspect, which delivers our national 'spring moment'. It's late April, the first evening warm enough to have tea with the backdoor open. The heavy smell of lilac is drifting in. A blackbird flies up to the roof – that confident swoop – and starts to sing, a song so reflective and relaxed that it captures the whole aura of the evening. If you are lucky, you will have one such moment every year of your life, and each time it will remind you of all the ones before. It would be nice to believe the blackbirds enjoy it, too.

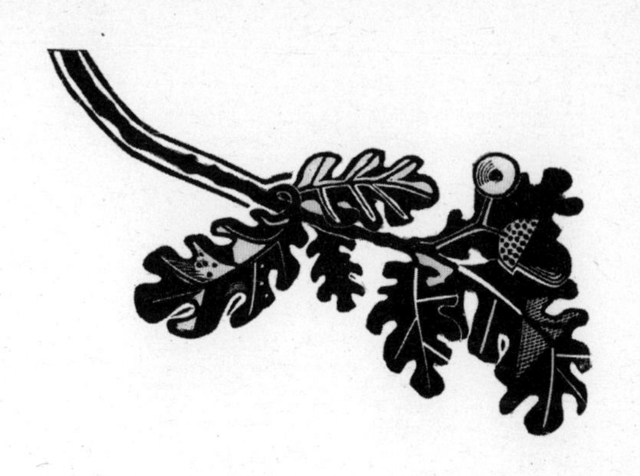

THE MAP AND THE WORD

6

The Map and the Word

DURING THE 1980S there was a flurry of scientific interest in one of the contenders for a sixth sense – the sense of direction. It looked for while like a promising bridge between the Romantic imagination and natural science. Was there something in ley lines beyond Druidic mysticism? Did dowsing work? Why could some people orientate themselves in space and others not?

Scientists from Manchester University came up with some intriguing experimental evidence. They found that dowsing – for metal pipes and underground rocks as well as subterranean water – worked; that most people had the ability to do it, and that a coat-hanger was just as good as the traditional and talismanic hazel fork. Another experiment showed that the majority of people could find their way in the general direction of home from an unfamiliar place, but not if their heads were encased in lead helmets. The conclusions were more speculative, but physiological tests suggested it was a group of muscles in the shoulders that made the twig or hanger twitch, not some mysterious link between wood and water, and that maybe these muscles were neurologically wired up to iron-rich cells in the bridge of the nose. It looked as if an anciently inherited sensitivity to small variations in the earth's magnetic field (known from other animals) might still persist in humans.

The research fizzled out after a few years, maybe because of too close links with New Age cults. But there's another route to knowing where you are that is beloved of Romantics. It's not a single sense in the physiological meaning of the word, but that confection of sight, smell, sound and memory that makes us register the *genius loci*, the spirit of place. Its technological companion is the map, whose impact – in common with all sense-extensions – is ambiguous.

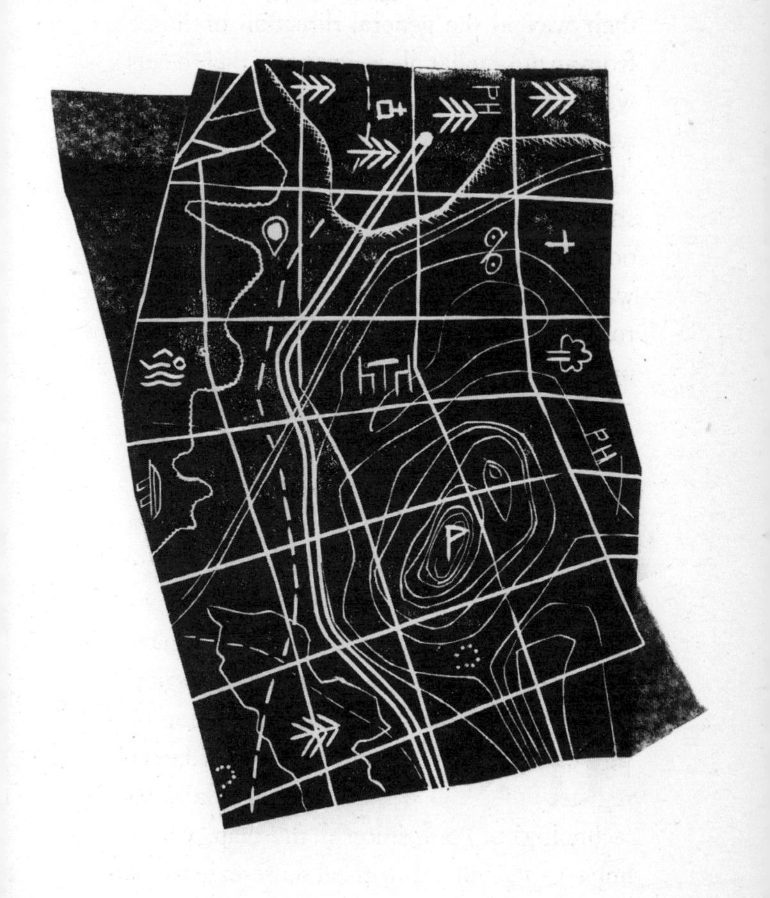
PH
PH
P

I've been a map-worm since I first began exploring the countryside. Maps, I discovered, weren't just aids to find your way about; they were a kind of cryptogram. Reading them intensely was an act of divination, which could liberate the *genius loci* from its coded ciphers. Maps were paper dreams.

When I lived in the Chilterns, it was the commons, not the human settlements, which seemed like the centres of civilisation on the map. Paths radiated from and through them like a starburst. When I first discovered East Anglia, I was captivated, in contrast, by the oriental delicacy of its mapped countryside. The outlines of tiny woods echoed the ebb and flow of ancient greenways. The village names were strung out along yellow B-roads like concrete poems. As I pored over the Ordnance Survey charts, I tried to imagine what these unvisited corners would be like in reality – the nuances of slope, the declensions of light and shade, where favourite plants would be growing – *exactly* where, since a diviner's pride was at stake. When I moved to Norfolk, recovering after a long illness, I'd read the entire local landscape before I set foot in it.

The valley I live in now is a complicated essay in the varieties and forms of wetness, and on the map the interconnected dykes and ponds and streamlets have the look of a sheet of splintered ice, a fragmented fenland still held together by the fluidity of water.

Some of this incidental poetry is apt to disappear when you visit places in the flesh. In the end maps *are* only ciphers. They were never intended to catch the character of the landscape. This isn't just because they are drawn from a point of view only available to us in an aeroplane. There are also problems of scale. Maps are chiefly intended for people on the move, so roads are shown as hugely inflated. They dominate both the physical area of a map (a trunk road would be hundreds of yards wide at the same scale as its mapped counterpart) and its imaginative reach. Marked roads and tracks suggest where you should go, the order in which you should view things. They are the modern map's centre of gravity, the skeleton which gives logic and structure to the entire landscape.

But there are other kinds of chart, less human-centred. I saw one after a rare snowfall some years ago, in a meadow I knew well in the Chilterns. I'd forgotten how dramatically snow redraws the map. Everything shallow and insubstantial in the landscape – roads, crops, surface soil – simply vanishes. All that remains are the gaunt fundamentals of hill, dip, rock and tree. But the snow itself then becomes a blank sheet for a new kind of plan that sketches out the ephemeral routeways and land-uses of myriad other creatures. On the meadow that morning there were the lurid stains of bird droppings under a rowan bush; the ancestral pathways that badgers followed out of their sett in the wood, full of curves and diversions and scuffled pauses; small mammal prints going round in circles. Footnotes to the events of the night.

The meadow that morning was a revelation, and I began to develop a taste for feral walking, deliberately avoiding human paths, and allowing the urge to potter to overcome any intention to get somewhere. I'm intrigued by what pulls me this way and that. Big trees on the horizon are an obvious magnet, but so are tiny self-sown oaklings, the beginnings of future woods. Any toadstool makes me tack, but so do other curious breaks in the surface: animal bones, anthills, wet hollows. Sometimes I try to navigate by animal tracks alone, but these routes are as wobbly as my own ambulatory doodles.

I'm not sure what kind of map I am building up when I potter like this. It's spontaneous and fleeting, but also, I've found, quite memorable. It gives objects in the landscape a quite different significance from those they have on a formally drawn map. Somehow, I guess, following your nose (or your inner eye) brings together a vestigial animal curiosity with your own cultural fascinations.

I think I must have moved about like this as a young boy, exploring that wasteland that lay at the back of our house. The Field felt like a second skin, prickly with sensations, full of meaning and associations, but never in a consciously reflective way. (Much later I learned that in parts of the Caroline Islands of the Pacific the inhabitants had plans of their home ground actually tattooed on their skin.) Now, up in the flatlands of Norfolk, the landscape is less emotionally charged, and I am consciously reflecting on its echoes.

Our house is only a few hundred yards from the edge of Norfolk's great arable plateau. There's scarcely a trace left of the great hornbeam forest that grew here until medieval times – a shocking thing for someone who did their growing-up in wood country. As for the farmed landscape, it's not so much flat as flattened. Developing a Romantic sense of place here isn't easy.

But in the wetland and fens in the valleys, it's another matter. They have a spirit of place that is mercurial – shiny but fleeting. You register it not so much from immobile markers in the landscape, as from ambience and rhythm, moments of natural theatre. The dialect of Norfolk's wetlands is made up from scribbles of geese across the sky, the bare fretwork of alder twigs in winter, the seductiveness of dykes disappearing into the reeds. Water animates a landscape, makes it impossible to pin down in the fixed geometry of a map. A minor flood up here turns dry hollows into lakes, and ferries the solid lumps of the landscape – fallen branches, fern clumps, roots, the soil itself – into new situations. Water is the connective tissue of these places, but it has none of the static rigidity of a road system on a map.

Wetness also gives the lie to the cliché that Norfolk is flat – or 'VERY flat' in the words of Noël Coward's notorious put-down. That cliché has always seemed to me a perversion of the truth – and not because some parts of north Norfolk are as hilly as Wiltshire. But I couldn't work out why, until I saw that Coward's quip – which has stereotyped Norfolk's landscape ever since – was, again, a map made from a peculiarly limited human viewpoint – about five feet something above the level of the ground. It was a literally superficial perspective, too – of the single dimension of surface elevation. Much of what is important to the life of fen or marsh depends on what is *under* the surface. It's here that the mosaic of pools, the alternation of vegetation, the long rhythms of succession, are determined. One day, maybe inspired by John Clare's poetic vision of a swamp from the point of view of a snipe, it occurred to me that you could re-imagine a wetland as a landscape whose contours were *under* the ground, as an inverted habitat, riddled with concave depressions. You could then imagine it turned upside

down. The prospect would suddenly erupt with mounds and banks. Glacial hollows and peat-pits dug out by humans would swell like prehistoric barrows. Dykes would appear like three-dimensional fences.

Then, in my mind, I turned it the normal way up again, and tried to see the surface of the marsh – the conventional map – as an outgrowth of this damp labyrinth below. The idea that it is a flat place, in the sense of being homogeneous, became ridiculous. I could make out dark sedges, livid bog-mosses, mist-green patches of reed. There are grass tussocks, scrubby tumps, flashes, pools, inscrutable ribbons of vegetation, a mosaic of tonal elevations. And the whole prospect is in constant motion. Even the birds and insects that float above the surface are like dowsers of this deep world, responding to invisible currents in the air, tiny thermals generated by the minute shifts from water to grass to reed.

Henry Thoreau had a vision of the true map-maker and would-be writer about nature as a kind of amanuensis for the wild: 'he would be a poet', he dreamed, 'who could impress the winds and streams into his service, to speak for him; who nailed words to their primitive senses, as farmers drive down stakes in the spring … who derives his words as often as he used them, transplanted them to his page with earth adhering to their roots; whose words were so true and fresh and natural that they would appear to expand like buds at the approach of spring.'

In one sense this is what John Clare did, not just in his insistent use of evocative dialect, but in the directness of his vision. He looked the world in the eye, on the level, in its own terms. But he knew his was a human perspective, and he spoke up for the human voice as well as for nature's. In a poem called 'Shadows of Taste' he repeats his hatred for the narrow scientific view of the collector and all those who 'steal nature from its proper dwelling place'; and applauds instead the man of taste – by which he means anyone who appreciates wild things in their context, which includes their accumulated cultural associations. But in a remarkable stanza, he insists that 'taste' is a faculty enjoyed by all living things. It's their unconscious, inherited choice of – and comfortableness in – their own habitats. Their internal map, if you like. By the startlingly original use of a single word he puts all beings on a common footing:

> Not mind alone the instinctive mood declares
> But birds & flowers & insects are its heirs
> Taste is their joyous heritage & they
> All choose for joy in a peculiar way

What I hope has emerged from these essays is how powerful our unassisted senses are when they're guided by the imagination. Scientific insight and technological enhancement can open new perspectives, but it's in our gift to use this to change our *ordinary* point of view. Language, that special human gift, itself becomes a transforming instrument. It can expand the imagination as arrestingly as an electron microscope. Clare sees taste as a faculty of all living things. Gary Snyder describes animal trails as narratives. Henry Thoreau creates a poetic analogy of botanical taxonomy. Marcel Proust glimpses the role of scent as a universal carrier of memory and message. Richard Dawkins interprets birdsong as emotional persuasion. Lewis Thomas has a vision of the earth, seen from space, as a single cell. All these remarkable insights begin with a specialised, near-scientific revelation, but are transformed into a common perspective by a leap of the imagination, through the ever-fresh air of language. The word becomes a map of the new world.